BIG YELLOW MACHINES

Excavators

Written by Jean Eick
Illustrated by Michael Sellner

ABDO & Daughters

PUBLISHING

BIG YELLOW MACHINES

Published by ABDO & Daughters Publishing
4940 Viking Drive, Suite 622
Edina, Minnesota 55435 USA

Designed by Michael Sellner
Edited by Jackie Taylor
Production: James Tower Media • Design
Photo Credits: "Images © 1995 PhotoDisc, Inc."
and Caterpillar Image Lab

Printed in the United States of America

Library of Congress Cataloging-in-Publication Data
Eick, Jean, 1947-
    Excavators / written by Jean Eick; illustrated by Michael Sellner.
       p.   cm. – (Big yellow machines)
Summary: Gives information about this machine by listing its working parts,
describing its cab, and explaining how it is used throughout the world and
how it is constructed.
    ISBN 1-56239-730-3
    1. Excavating machinery – Juvenile literature. [1. Excavating machinery.
2. Machinery.] I. Sellner, Michael, ill. II. Title. III. Series :Eick, Jean, 1947-Big
Yellow Machines.
TJ1355.E35   1996   96-14128
624. 1'52—dc20   CIP
                                                                            AC

Contents

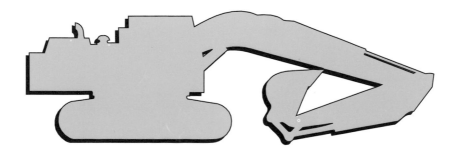

What is an  ?

Almost every place you see highway construction workers, you will see an excavator. Some people just call this machine a big digger. That's what an excavator does. It digs big holes, but it can do a lot more than that. An excavator can also lift, carry and dump.

It's fun to watch an excavator at work! The giant arm bends down and the bucket digs into the ground. When the bucket is full, the arm comes up and the machine swings around. Then, with a clunk, it drops its load of dirt. Before the dust settles, the bucket is back in the ground, ready for another load.

Uses for an Excavator

An excavator is used to dig the foundation of a building. A foundation is a solid base for the building to sit on. The building would fall over or sink into the ground without it.

This house is built on a foundation.

Excavators can also make trenches. A trench is a long hole used for pipes. The pipes can carry electricity, water and gas.

Excavators help build roads. Highway crews can use excavators to dig ditches or remove giant rocks. They can also use them to fill trucks with gravel or dirt.

Forestry crews use excavators, too. They use them on rocky mountains and deep in the forest. The crews like excavators because they can do so many things.

Excavators are really put to work in mines and quarries. Miners use the giant front shovels to dig rocks and dirt. They even remove sand and gravel to mine gold.

At work in a mine.

Excavators do important work along rivers and lakes. They can pile up rocks in rivers and streams to make dams. They can clear the banks using a special tool called a river rig. Thanks to excavators, we are able to rebuild many streams and rivers.

Clearing the bank with a river rig.

Inside the Cab

The cab is just like an office for the operator. There are windows all around so the operator can get a good view in every direction. Thick floor mats help to keep the cab quiet inside and a special monitor even tells

the operator important information about the fuel, oil and engine. Everything the operator needs to control the excavator is right inside the cab. Levers and pedals make the machine move and turn. Joysticks are used to control the bucket and move the long arm in different directions.

Lever

Pedals

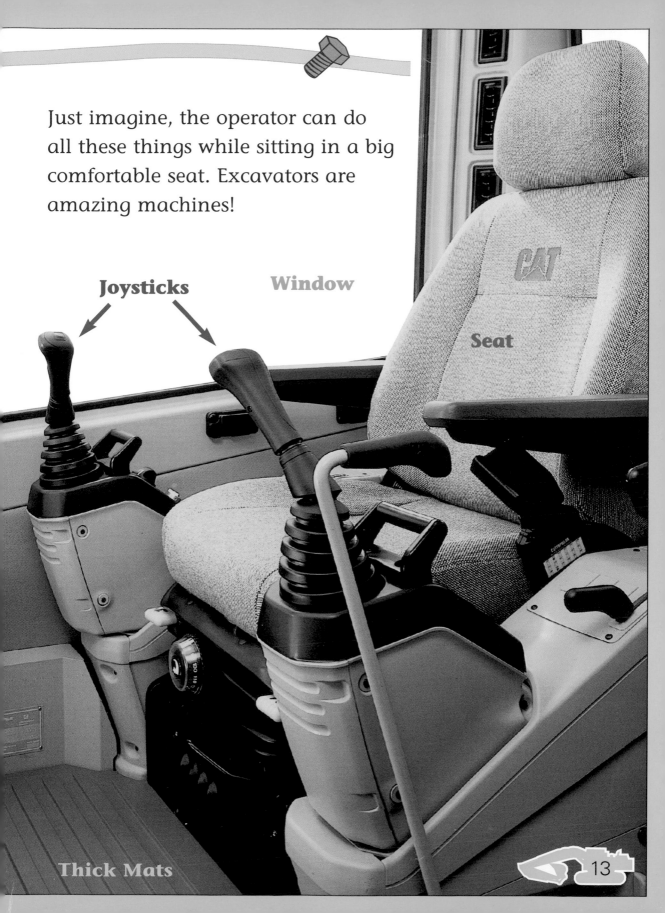

Just imagine, the operator can do all these things while sitting in a big comfortable seat. Excavators are amazing machines!

Joysticks

Window

Seat

Thick Mats

Hydraulic Ram - Moves the stick in and out.

Boom - The long part of the arm that is attached to the upper part of the excavator.

Hydraulic Hoses - Oil is pumped through these hoses to run parts of the machine.

Cab - Where the operator sits to run the excavator.

Engine - Where the power comes from to run the machine.

Track Shoes - The individual parts of the crawler tracks.

Undercarriage - The work platform which includes the crawler tracks and track roller frames.

Parts of an Excavator

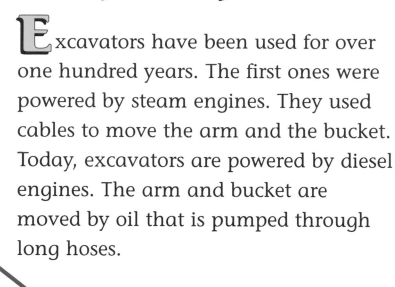

Excavators have been used for over one hundred years. The first ones were powered by steam engines. They used cables to move the arm and the bucket. Today, excavators are powered by diesel engines. The arm and bucket are moved by oil that is pumped through long hoses.

Stick - The end of the arm that holds the bucket or work tool.

Bucket - The shovel or scoop that digs into the ground.

Crawler Tracks - Huge belts that take the place of tires, moving the machine as they spin slowly around the frame.

An early excavator at work.

Tools of an Excavator

Excavators can do many other important jobs when special tools are added to the stick. Look at all the things they can do!

Mower

Tilting Buckets

Quick
Coupler

Clamshell
Bucket

Hydraulic Hammer

Material Handling Stick

Where in the world can you find an Excavator?

Excavators are found in almost every country in the world. They are used to mine gold in Ecuador and build airports in Hong Kong. They work on quiet streets in France and busy highways across the United States. Excavators are important machines all around the world.

North America

United States

Ecuador

South America

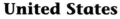

United States

Ecuador

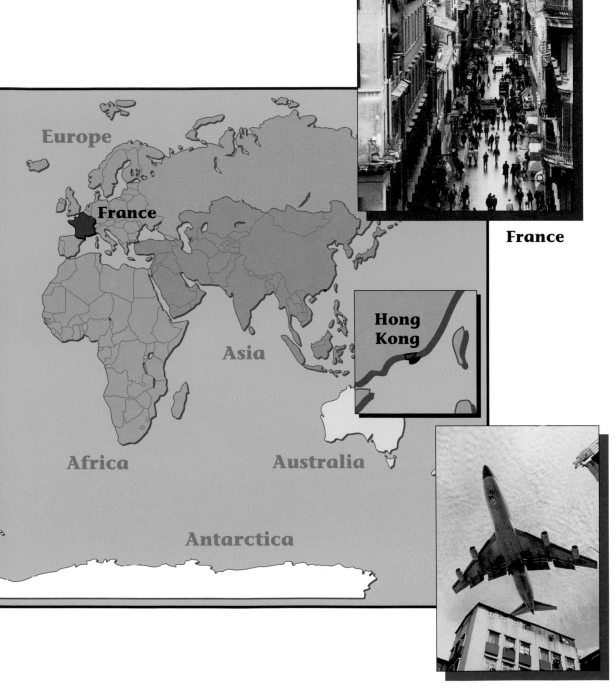

Europe

France

Asia

Africa

Australia

Antarctica

Hong
Kong

France

Hong Kong

Where in the world can you find an Excavator?

Indonesia

Excavators are used in Indonesia to drill for natural gas, and mine for gold and tin. People in Indonesia need more roads and buildings. Excavators can sure help out there!

Where in the world can you find an Excavator?

South Africa

325 L CAT

South Africa is a country that is making many changes. Like Indonesia, it needs more roads, buildings and machines for mining. Excavators can help South Africa grow and change.

Where in the world can you find an

Excavator?

CHINA

Coal mining is important in China, so a lot of excavators are needed there.

Excavators are also important in China because they provide many jobs for the people who make them.

These men are getting the excavator ready for a job.

Where in the world can you find an Excavator?

Excavators really do help to make the world a better place. When Hurricane Andrew damaged many homes and buildings in Florida, they were used to help clean up the mess. In Russia, excavators were used to take apart dangerous missiles, airplanes and submarines. In Kuwait, they were used to stop fires during the Gulf War. Excavators even helped take down the Berlin Wall so that people in Germany could be friends again.

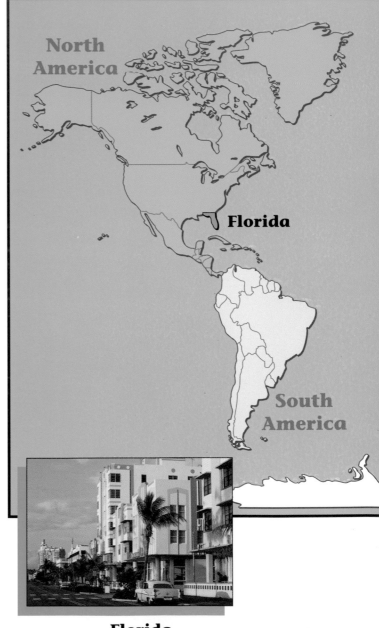

North America

Florida

South America

Florida

Excavators have been used around the world to keep people safe and happy.

Europe

Russia

Germany

Kuwait

Japan

Asia

Africa

Australia

Antarctica

Japan

Russia

Kuwait

27

Special Facts About
Excavators

Excavators come in different sizes. Even the smallest ones are very big and strong, but it's the largest ones that are the most amazing. They are gigantic!

How Much Do Excavators Weigh?

The small excavator weighs as much as a full-grown elephant (14,000 pounds/6,600 kilograms).

The gigantic excavator weighs more than the blue whale, the biggest animal on Earth.

The blue whale can weigh up to 300,000 pounds (136,000 kilograms). The gigantic excavator weighs 370,000 pounds (168,000 kilograms)!

How Much Fuel Do Excavators Hold?

The small excavator holds about as much fuel as two cars (29 gallons/110 liters).

The gigantic excavator holds 687 gallons (2,600 liters). That much fuel could fill 11 school buses!

Small Excavator

🔩 How Deep Can Excavators Dig?

The small excavator can dig a hole 15 feet 3 inches (4.6 meters). A full-grown giraffe standing in a hole that deep could barely see out.

The gigantic excavator can dig a hole 27 feet 6 inches (8.4 meters). A giraffe would need a ladder to see out of the top of that hole!

🔩 How Much Can Excavators Lift?

The small excavator can lift 10,915 pounds (4.954 kilograms). That's more than a full-size rhinoceros in the front bucket.

The gigantic excavator can lift 52,800 pounds (23,950 kilograms). That's as much as two rhinos, an elephant and a killer whale all at the same time. Imagine that!

Gigantic Excavator

Words To Remember

Cab - Where the operator sits to run the machine.

Engine - Where the power comes from to run the machine.

Excavator - A rugged, hardworking machine that can dig, lift and carry.

Foundation - The solid base that keeps a building from sinking into the ground.

Joystick - Levers inside the cab used by the operator to control the bucket and boom on the excavator.

Trenches - Long, narrow holes used to hold pipes.

Index